Stepping into Standards Theme Series

Nutrition

Contributing Writers
Kim Cernek
Adela Garcia
Pamela Jennett
Starin Lewis
Marilyn Marks
Heather Phillips
Vicky Shiotsu

Editor: Sheri Rous
Illustrator: Jenny Campbell
Cover Illustrators: Darcy Tom and Kimberly Schamber
Designer: Moonhee Pak
Cover Designer: Moonhee Pak
Art Director: Tom Cochrane
Project Director: Carolea Williams

© 2003 Creative Teaching Press, Inc., Huntington Beach, CA 92649
Reproduction of activities in any manner for use in the classroom and not for commercial sale is permissible.
Reproduction of these materials for an entire school or for a school system is strictly prohibited.

Table of Contents

INTRODUCTION 3

GETTING STARTED

How to Use This Book 4
Meeting Standards 6
Introducing . . . Healthy Nutrition 8
Nutrition Pretest 9
What Do You Know? KWL Chart 10
Food Dude to the Rescue—Reading Comprehension Test 11

LANGUAGE ARTS

What's the Food Guide Pyramid? 13
Nutrition Word Break 14
Digestion Sequence 15
Vitamins and Minerals 16
Food Label Scavenger Hunt 17
Hooray for a Healthy Meal! 18
Watch What You Eat 18
Take a Bite 19
Five Senses Poetry 19
Food Guide Pyramid Advertisement 20
Sweet Dream Pie 21
Nutrition Mix-Up 21

MATH

Fruity Favorites 29
Marvelous Milk 30
How Much in a Can? 31
Grocery Store Challenge 32
Terrific Trail Mix 33
Calorie Count 34
Food Pyramid Pizza 35

SCIENCE

Why Do I Need Them? 42
The Pyramid of Health 43
Water for Life 44
Collecting Water 45
The Water in Food 46
Fats in Foods 47
Breaking Apart Fats and Oils 48
Vitamin C Is Good for Me 49

SOCIAL STUDIES

Comparing Kitchens 52
National Nutrition 52
Graphing What to Grow 53
Good to Grow 54
Wants and Needs 55
Food Election 56
Goods and Services 57

EATING RIGHT CUMULATIVE TEST 62
CERTIFICATE OF COMPLETION 64

Introduction

Due to the often-changing national, state, and district standards, it is frequently difficult to "squeeze in" fascinating topics for student enrichment on top of meeting required standards and including a balanced program in your classroom curriculum. The *Stepping into Standards Theme Series* incorporates required subjects and skills for second- and third-grade students while engaging them in an exciting and meaningful theme. Students will participate in a variety of language arts experiences to help them with **reading** and **writing** skills. They will also enjoy **standards-based math activities, hands-on science projects,** and **interactive social studies activities.**

The creative lessons in *Nutrition* provide imaginative, innovative ideas to help you motivate students as they learn about healthy nutrition in your classroom. The activities will inspire students to explore healthy eating as well as provide them with opportunities to enhance their knowledge and meet state standards. The pretest and posttest will help you assess your students' knowledge of the subject matter and skills.

Invite students to explore nutrition as they
- learn about the different food groups that make up the Food Guide Pyramid
- read and write healthy recipes
- measure ingredients
- sequence the steps to digestion
- learn the components of food labels
- discover our bodies are made up of 70% water
- compare kitchens of the past and present
- mark locations on a topographical map to determine the best location to grow crops

Each resource book in the *Stepping into Standards Theme Series* includes standards information, numerous activities, easy-to-use reproducibles, and a full-color overhead transparency to help you integrate a fun theme into your required curriculum. You will see how easy it can be to incorporate creative activities with academic requirements while students enjoy their exploration of healthy nutrition!

Getting Started

How to Use This Book

This comprehensive resource is filled with all the components you need to introduce, teach, review, and assess students on key skills while still making their learning experience as memorable as possible. The lessons are divided into four main sections: Language Arts, Math, Science, and Social Studies. Follow these simple steps to maximize student learning.

1. Use the **Meeting Standards** chart (pages 6–7) to help you identify the standards featured in each activity and incorporate them into your curriculum.

2. Review **Introducing . . . Healthy Nutrition** (page 8). This page provides numerous facts about the theme of study, literature selections that work well with the theme, key vocabulary words that your students will encounter while studying the theme, and the answers to all the assessments presented throughout the resource. Use this page to obtain background knowledge and ideas to help you make this a theme to remember!

3. Use the **Nutrition Pretest** (page 9) to assess your students' prior knowledge of the theme. This short, knowledge-based, multiple-choice test focuses on the key components of healthy nutrition. Use the results to help determine how much introduction to provide for the theme. The test can also be administered again at the end of the unit of study to see how much students have learned.

4. Copy the **What Do You Know? KWL Chart** (page 10) onto an overhead transparency, or enlarge it onto a piece of chart paper. Ask students to share what they already know about healthy nutrition. Record student responses in the "What We **Know**" column. Ask students to share what they would like to know about healthy nutrition. Record student responses in the "What We **Want to Know**" column. Then, set aside the chart. Revisit it at the end of the unit. Ask students to share what they learned about nutrition, and record their responses in the "What We **Learned**" column.

5 Give students the **Food Dude to the Rescue—Reading Comprehension Test** (pages 11–12). It is a great way to introduce students to the theme while making learning interesting. You can assess your students' comprehension skills as well as introduce students to the components of healthy nutrition. The multiple-choice questions require students to use literal as well as inferential skills.

6 Use the **Food Guide Pyramid full-color transparency** to enhance the theme. Display the transparency at any time during the unit to support the lessons and activities and to help reinforce key concepts about healthy nutrition.

7 Use the activities from the **Language Arts, Math, Science, and Social Studies sections** (pages 13–61) to teach students about healthy nutrition and to help them learn, practice, and review the required standards for their grade level. Each activity includes a list of objectives, a materials list, and a set of easy-to-follow directions. Either complete each section in its entirety before continuing on to the next section, or mix and match activities from each section.

8 Use the skills-based **Eating Right Cumulative Test** (pages 62–63) to help you assess both what your students learned about the theme and what skills they acquired while studying the theme. It will also help you identify if students are able to apply learned skills to different situations. This cumulative test includes both multiple-choice questions and short-answer questions to provide a well-rounded assessment of your students' knowledge.

9 Upon completion of the unit, reward your students for their accomplishments with the **Certificate of Completion** (page 64). Students are sure to be eager to share their knowledge and certificate with family and friends.

Meeting Standards

Language Arts

	What's the Food Guide Pyramid? (PAGE 13)	Nutrition Word Break (PAGE 14)	Digestion Sequence (PAGE 15)	Vitamins and Minerals (PAGE 16)	Food Label Scavenger Hunt (PAGE 17)	Hooray for a Healthy Meal! (PAGE 18)	Watch What You Eat (PAGE 18)	Take a Bite (PAGE 19)	Five Senses Poetry (PAGE 19)	Food Guide Pyramid Advertisement (PAGE 20)	Sweet Dream Pie (PAGE 21)	Nutrition Mix-Up (PAGE 21)
READING												
Comprehension	•		•	•							•	
Interpret Information			•	•	•							
Sequencing			•									
Syllabication		•										
Vocabulary Development	•											
WRITING												
Adjectives						•			•	•		
Alphabetical Order												•
Editing										•		
Friendly Letters						•						
Nouns							•					
Penmanship						•			•	•		
Poetry									•			
Recipes											•	
Revision										•		
Sentences						•	•		•	•		
Spelling		•				•			•			•
Verbs							•					

6 Getting Started

Meeting Standards

Math · Science · Social Studies

	Fruity Favorites (p.29)	Marvelous Milk (p.30)	How Much in a Can? (p.31)	Grocery Store Challenge (p.32)	Terrific Trail Mix (p.33)	Calorie Count (p.34)	Food Pyramid Pizza (p.35)	Why Do I Need Them? (p.42)	The Pyramid of Health (p.43)	Water for Life (p.44)	Collecting Water (p.45)	The Water in Food (p.46)	Fats in Foods (p.47)	Breaking Apart Fats and Oils (p.48)	Vitamin C Is Good for Me (p.49)	Comparing Kitchens (p.52)	National Nutrition (p.52)	Graphing What to Grow (p.53)	Good to Grow (p.54)	Wants and Needs (p.55)	Food Election (p.56)	Goods and Services (p.57)
MATH																						
Data Analysis and Probability	•					•																
Fractions			•		•		•															
Graphing									•													
Measurement			•		•		•															
Number and Operations	•	•	•	•		•	•															
Problem Solving		•	•																			
Reasoning and Proof		•																				
SCIENCE																						
The Characteristics of Organisms								•	•	•	•			•	•							
Investigation & Experimentation								•	•	•	•	•	•	•	•							
Organisms & Their Environments											•	•										
Properties of Earth Materials											•											
SOCIAL STUDIES																						
Ancestors																	•					
Community																						•
Economics																	•		•	•		•
Government																	•				•	
Mapping																			•			

Getting Started 7

Introducing... Healthy Nutrition

FACTS ABOUT NUTRITION

- The Food Guide Pyramid is a basic outline of how much food to eat from each of the six food groups.
- Foods are organized into groups according to the nutrients they contain.
- Foods give us energy and nutrients.
- It is important to eat foods from the Extras group sparingly because they contain few nutrients.
- All foods give us energy (calories) to keep our bodies working and moving every day.
- It is important for people to eat foods with plenty of vitamins and minerals in order to maintain good health.
- Eating foods that contain many vitamins and minerals, drinking plenty of water, and getting a good night's sleep will help a person stay healthy.
- During a person's lifetime, he or she will eat approximately 60,000 pounds (27,240 kg) of food.

LITERATURE LINKS

Eat Healthy, Feel Great by William Sears (Little, Brown & Company)

Eat Your Vegetables! Drink Your Milk! by Dr. Alvin Silverstein, Virginia Silverstein, and Laura Silverstein Nunn (Scholastic)

The Edible Pyramid: Good Eating Every Day by Loreen Leedy (Holiday House)

The Food Pyramid by Joan Kalbacken (Scholastic)

From Cow to Ice Cream: A Photo Essay by Bertram T. Knight (Scholastic)

Good Enough to Eat: A Kid's Guide to Food and Nutrition by Lizzy Rockwell (HarperCollins)

Long Ago and Today by Rozanne Lanczak Williams (Creative Teaching Press)

The Milk Makers by Gail Gibbons (Simon & Schuster)

Sweet Dream Pie by Audrey Wood (Scholastic)

What Happens to a Hamburger? by Paul Showers (HarperCollins)

Why Do People Eat? by Kate Needham (EDCP)

VOCABULARY

calcium

carbohydrates

energy

fats

minerals

nutrients

protein

vitamins

ASSESSMENT ANSWERS

Nutrition Pretest (PAGE 9)
1. *d* 2. *c* 3. *d* 4. *a* 5. *c* 6. *b* 7. *d* 8. *b*

Food Dude to the Rescue—Reading Comprehension Test (PAGES 11–12)
1. *b* 2. *d* 3. *a* 4. *d* 5. *c* 6. *d*

Eating Right Cumulative Test (PAGES 62–63)
1. *c* 2. *b* 3. *c* 4. *a* 5. *b* 6. *d* 7. *a* 8. *c* 9. *b* 10. *d* 11. *Answers may vary*
12. *Answers may vary*

Name_____   Date_____

Nutrition Pretest

Directions: Fill in the best answer for each question.

1 How many food groups can be found in the Food Guide Pyramid?
- (a) 3
- (b) 4
- (c) 5
- (d) 6

2 Which of these foods contains only a few nutrients?
- (a) chicken
- (b) carrots
- (c) cake
- (d) apples

3 Which food group should you eat from sparingly?
- (a) Meats, Beans, & Nuts
- (b) Milk & Milk Products
- (c) Breads & Grains
- (d) Extras

4 True or False: You should eat many different foods from each food group every day.
- (a) true
- (b) false

5 Which of the following is **not** something you should do every day to stay healthy?
- (a) get a good night's sleep
- (b) drink plenty of water
- (c) drink plenty of soda
- (d) eat a variety of foods from each food group

6 What else do foods give us besides energy?
- (a) a stomachache
- (b) nutrients
- (c) a treat
- (d) sleep

7 Which of these foods would fit in the Breads & Grains Group?
- (a) an apple
- (b) pudding
- (c) an egg
- (d) pizza

8 Which of these would be a healthy food choice?
- (a) gravy
- (b) tacos
- (c) French fries
- (d) chocolate bar

Getting Started **9**

What Do You Know? KWL Chart

What We Know	What We Want to Know	What We Learned

Name_____ Date_____

Food Dude to the Rescue—
Reading Comprehension Test

Directions: Read the story and then answer the six questions.

It was almost lunchtime and Sally and her brother Tom were having trouble deciding what they wanted to eat. They both agreed that they wanted to eat a healthy meal, but they could not remember which foods were in each food group.

All of a sudden, Sally and Tom heard a big "bang" and before their eyes appeared Food Dude. Sally remembered Food Dude from school. He starred in a video about eating healthy that her teacher had shown the class.

"It seems you both are having a hard time trying to figure out a healthy meal to eat. Mind if I help?" asked Food Dude. Within seconds, Food Dude pulled his Food Guide Pyramid poster out of his pocket and began sharing details about the different food groups. He explained how much of each serving a person should eat daily from each group.

"The Breads & Grains Group is the largest group. A person should eat 6–11 servings from this group each day. A person should eat 3–5 servings from the Vegetables Group and 2–4 servings from the Fruits Group each day. Don't forget to eat 2–3 servings from both the Milk & Milk Products Group and the Meats, Beans, & Nuts Group. The foods found in these groups contain many vitamins and minerals that give you lots of energy and help you stay healthy. A person should eat a variety of foods from each group every day. Eat foods from the Extras group, such as sweets, fats, and oils, sparingly because they contain few nutrients. It is also important to drink plenty of water," explained Food Dude.

Food Dude recommended that Sally and Tom eat a chicken sandwich with no mayonnaise and an apple for lunch and drink a glass of milk. He also told them that it is important to get a good night's sleep and exercise every day. Sally and Tom thanked Food Dude for his help and invited him to stay for lunch. Sally and Tom were excited to share with their parents everything they learned about eating healthy.

Name_____   Date_____

Food Dude to the Rescue—
Reading Comprehension Test

1 Why is the Food Guide Pyramid important?

ⓐ It explains how to make a pyramid.

ⓑ It shows how much food to eat from each of the six food groups.

ⓒ It explains how to stack food into the shape of a pyramid.

ⓓ none of the above

2 From which two food groups is a person supposed to eat 2–3 servings each day?

ⓐ Fruits Group and Vegetables Group

ⓑ Breads & Grains Group and Milk & Milk Products Group

ⓒ Meats, Beans, & Nuts Group and Extras Group

ⓓ Milk & Milk Products Group and Meats, Beans, & Nuts Group

3 Which food group contains the fewest nutrients?

ⓐ Extras

ⓑ Breads & Grains

ⓒ Vegetables

ⓓ Fruits

4 In addition to getting a good night's sleep and drinking plenty of water, a person should . . .

ⓐ stay up late.

ⓑ drink plenty of soda.

ⓒ eat plenty of candy.

ⓓ eat a balanced diet.

5 Which two food groups were missing from Sally and Tom's lunch?

ⓐ Breads & Grains and Extras

ⓑ Fruits and Vegetables

ⓒ Vegetables and Extras

ⓓ Milk & Milk Products and Extras

6 What should a person do in order to stay healthy?

ⓐ Get a good night's sleep.

ⓑ Eat a variety of foods from different food groups every day.

ⓒ Exercise.

ⓓ all of the above

12 Getting Started

Language Arts

What's the Food Guide Pyramid?

OBJECTIVES

Students will
- apply knowledge from a story to answer questions about nutrition.
- use food terminology to develop vocabulary.

MATERIALS

- *The Edible Pyramid* by Loreen Leedy
- Food Guide Pyramid reproducible (page 22)
- Food Guide Pyramid transparency
- overhead projector

Read *The Edible Pyramid* to the class, and discuss the different food groups the animals learned about while visiting the pyramid-shaped restaurant. Encourage students to recall some of the balanced meals the maitre d' suggested for the guests. Record student responses on the board. Give each student a Food Guide Pyramid reproducible. Display the Food Guide Pyramid transparency. Ask students *What food group would a taco fit in?* Explain to the class that many foods fit into more than one food group and that a taco is a perfect example of this. Ask the class to identify what foods are in a taco (e.g., shell, meat, cheese, lettuce, salsa). Record student responses on the board. Ask volunteers to share which food group each part of the taco belongs to. Record student responses on the board. Ask students to share other examples of foods that represent many different food groups (e.g., pizza, macaroni and cheese, hamburgers). Record student responses on the board. Divide the class into pairs. Have partners select three foods from the board and list on a piece of paper the ingredients of each food. Ask partners to determine which food group each ingredient belongs to. Invite partners to share their findings with the class.

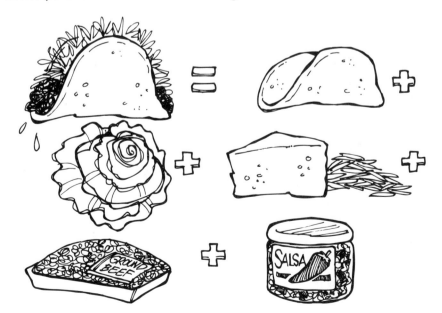

Nutrition Word Break

MATERIALS
- paper scraps
- cup

OBJECTIVES

Students will
- apply knowledge of basic syllabication rules.
- use prior experiences to help them spell words of varying degrees of difficulty.

Write on separate pieces of scrap paper the words from the word bank shown below, fold each paper in half, and place the folded papers in a cup. Explain to students that they will be playing a game called Nutrition Word Break. The object of the game is to score more points than the other team by correctly spelling the chosen word or stating the number of syllables the word has. Explain that each word is worth a specific number of points depending on the difficulty of the word (e.g., bread—2). If a team either misspells the word or incorrectly identifies the number of syllables in it, then the team does not get any points. Divide the class into two teams. Draw a paper from the cup, and say the word aloud. Invite the first team to spell the word or say how many syllables it has. Record the team's points on the board. Pick another word, and have the other team repeat the process. Continue with the remaining words. Have the teams add up their points to determine the winning team.

WORD BANK

bread—2	Food Guide Pyramid—5
grains—2	exercise—4
dairy—2	vegetables—4
carbohydrates—5	fruits—2
nutrients—5	grains—2
digestion—4	pasta—2
vitamins—3	calcium—4
minerals—3	stomach—4
moderation—5	

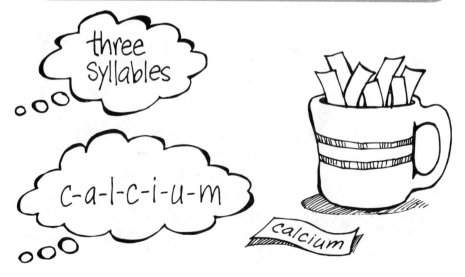

14 Language Arts

Digestion Sequence

OBJECTIVE
- Students will sequence the order in which food is digested.

MATERIALS
- *What Happens to a Hamburger?* by Paul Showers
- Steps to Digestion reproducible (page 23)
- scissors
- drawing paper
- glue

Read *What Happens to a Hamburger?* to the class. Explain to students that when they eat food, it passes through their body and is digested. Tell students that digestion is the way a person's body absorbs nutrients from food. Explain that they are going to learn about the steps of digestion but that in order to do this, you need their help to place the steps in the correct sequence. Give each student a Steps to Digestion reproducible. Invite volunteers to read each sentence. Have students cut apart the boxes and place the sentences in the correct order. Invite volunteers to share their sequence of digestion. Ask the class to verify that the steps are in the correct order. Give each student a piece of drawing paper. Have students glue their paper strips on the drawing paper in the correct order. Students should indicate that the steps for digestion are as follows:

You grind the food in your mouth and mix it with your saliva.

The esophagus in your throat pushes the food down to your stomach.

Your stomach mixes the food with acid until the food is a thick liquid.

The small intestine takes the thick liquid and absorbs the nutrients through its lining.

Nutrients enter the bloodstream through the small intestine's lining.

Then, the larger intestine takes over.

The large intestine absorbs water and passes on the unused parts.

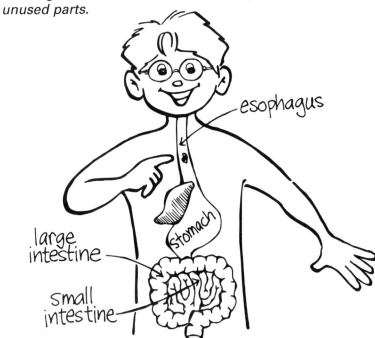

Vitamins and Minerals

MATERIALS

- *Eat Your Vegetables! Drink Your Milk!* by Dr. Alvin Silverstein, Virginia Silverstein, and Laura Silverstein Nunn
- drawing paper
- crayons or markers

OBJECTIVE

- Students will interpret information from a chart.

Explain to the class that food contains vitamins and minerals that help a person's body stay healthy. Read *Eat Your Vegetables! Drink Your Milk!* to the class, and discuss the components of a healthy diet. Divide the class into pairs. Assign each pair of students one of the following vitamins or minerals: vitamin A, B vitamins (B1, B2, B3, B6, B12), vitamin C, vitamin D, vitamin E, vitamin K, calcium, potassium chloride, iron, phosphorus, and sodium. Place *Eat Your Vegetables! Drink Your Milk!* at the back of the room, and open it to pages 20–21. Send one pair of students at a time to the back of the room. Tell students to find their vitamin or mineral on the chart in the book and read the information about it. Then, have pairs create a poster advertising the benefits of their vitamin or mineral. To challenge students, have them also include foods that contain their vitamin or mineral and what their vitamin or mineral does for the body. Have pairs share their completed poster with the class.

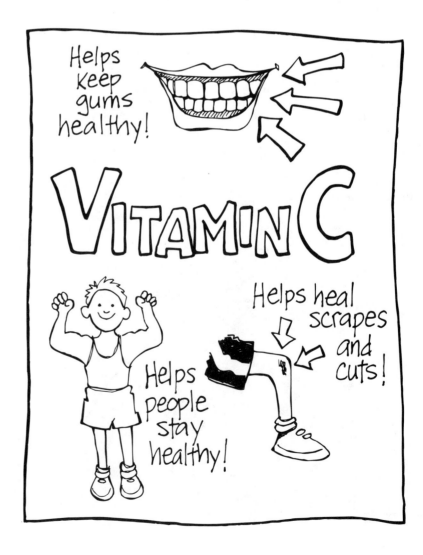

Food Label Scavenger Hunt

OBJECTIVE

- Students will interpret information from a chart.

MATERIALS

- Food Label Scavenger Hunt reproducible (page 24)
- cans or boxes of food

Ask volunteers to bring to class a box or can of food. Explain that they will get it back unopened to take back home once the activity is completed. Tell students that they will be going on a food label scavenger hunt. Explain that each box or can of food contains a nutrition label that lists the number of calories per serving, the amount and types of vitamins and minerals, the percentage of the recommended daily amount of vitamins and minerals, and the ingredients in the food. Display the unopened food items around the classroom. Divide the class into pairs. Give each pair of students a Food Label Scavenger Hunt reproducible. Tell students they will read the labels on the different food items to try to find a food item that matches one of the descriptions on the reproducible. Once they find a match, have them write the name of the food next to the description. For example, if a can of string beans has 0 grams of fat, students would write *string beans* next to *Find a food with 0 grams of total fat*. End the activity by calling the class back together and inviting students to compare answers. See if any pairs were able to find a matching food item for every description.

Language Arts

MATERIALS

- writing paper

Hooray for a Healthy Meal!

OBJECTIVES

Students will
- use adjectives to describe food items.
- write a friendly letter.

Encourage students to close their eyes and imagine that they have just eaten a healthy lunch at a restaurant. Have volunteers use descriptive words to help describe their lunch to the class. Record student responses on the board. Explain to the class that they will be writing a friendly letter to the owner of their make-believe restaurant to thank him or her for the delicious, nutritious meal. Review with students the parts of a friendly letter, including the date, salutation, body, closing, and signature. Encourage students to name in their letter specific healthy foods that they imagined eating and what food group each food belongs to. Invite volunteers to read their letter to the class.

MATERIALS

- Watch What You Eat reproducible (page 25)
- crayons or markers
- bookbinding materials

Watch What You Eat

OBJECTIVE

- Students will write declarative, interrogative, imperative, and exclamatory sentences.

Review with students the four types of sentences: declarative, interrogative, imperative, and exclamatory. Give each student a Watch What You Eat reproducible, and encourage students to discuss the pictures on the page. Invite students to imagine what the people in the four boxes are saying. Encourage students to write sentences in the speech bubbles to tell a story about nutrition. Challenge them to write one type of sentence in each bubble. Invite students to color the pictures. Bind the completed papers in a book titled *Watch What You Eat*.

18 Language Arts

Take a Bite

MATERIALS
- chart paper
- drawing paper
- crayons or markers

OBJECTIVE
- Students will identify nouns and verbs and use them correctly in sentences.

Draw a line down the center of a piece of chart paper. Title the left column *Nouns* and the right column *Verbs*. Review with students that nouns name a person, place, thing, or idea. Explain that verbs are "action" words. Invite students to name foods they like, and write them under "Nouns." Ask students to name words that describe ways to eat (e.g., nibble, bite, chew, swallow), and write them under "Verbs." Give each student a piece of drawing paper. Ask students to choose one noun and one verb from the chart, use them in a complete sentence, and write it at the bottom of their paper. Have students underline the noun and circle the verb they chose. Invite students to illustrate their sentence. Have students trade papers with a partner and say aloud the nouns and verbs in the sentences.

Five Senses Poetry

MATERIALS
- Five Senses Poetry reproducible (page 26)
- apple
- crayons or markers

OBJECTIVE
- Students will write a description that uses concrete sensory details to present and support a unified impression of a food.

Brainstorm with the class different nutritious foods. Record student responses on the board. Discuss with the class what makes these foods nutritious. Explain that our five senses are seeing, hearing, smelling, tasting, and touching. Invite a volunteer to come up to the front of the class. Have the volunteer close his or her eyes and touch an apple. Ask the volunteer to describe what the apple feels like. Then, tell the volunteer to open his or her eyes and describe how the apple smells, how it tastes, and if anything could be heard when a bite is taken out of it. For example, an apple looks colorful, smells tart, tastes sour, feels smooth, and sounds crunchy. Choose one food from the list, and have different volunteers describe it using their five senses. Give each student a Five Senses Poetry reproducible. Have students choose one of the foods from the list and use their senses to describe it. Have them complete the frames on the reproducible to write their poem. Invite students to illustrate their completed poem and then share it with the class.

Language Arts 19

MATERIALS

- *The Food Pyramid* by Joan Kalbacken
- Food Guide Pyramid transparency
- overhead projector
- food advertisements
- drawing paper
- crayons or markers

Food Guide Pyramid Advertisement

OBJECTIVE
- Students will use descriptive writing to write an advertisement.

Read aloud *The Food Pyramid*. Point out the different nutrients in various foods. Display the Food Guide Pyramid transparency. Review with the class the different food groups found in the Food Guide Pyramid, and discuss how many servings should be consumed from each group per day. Discuss the nutrients that are found in some of the foods in each group and how these nutrients are important to our health. For example, point out that vitamin C is good for our immune system. Tell students they will create an advertisement to try to get people to buy their chosen food. Explain that they need to use descriptive words and sensory details in their advertisement to help convince people to buy their food. Share food advertisements with the class. Have each student select a food in any category except the "Extras." Have students revise and edit their advertisement and then rewrite it on a piece of drawing paper. Encourage students to include an illustration that is designed to catch the audience's attention and promote the product.

Sweet Dream Pie

MATERIALS
- *Sweet Dream Pie* by Audrey Wood
- Nutritious Pie Recipe reproducible (page 27)
- recipe books

OBJECTIVES

Students will
- read a recipe.
- write their own recipe.

Read aloud *Sweet Dream Pie,* and discuss with students why the characters on Willobee Street had difficulty sleeping. Students should say that the characters ate too many sweets. Lead the class in a discussion about what would really happen if people ate all the time like the characters in the story. Have students brainstorm the negative side effects of a sugar-filled diet. Discuss some foods students like to eat that are nutritious. Share examples of recipe books, and discuss the format of the recipes. For example, point out the list of ingredients and the cooking directions. Model for the class how to write a recipe for a nutritious dish. Give each student a Nutritious Pie Recipe reproducible. Have students write a recipe for a "nutritious pie" using numerous nutritious foods they like to eat. Invite students to share their recipe with the class. Display the creative, nutritious recipes on a bulletin board.

Nutrition Mix-Up

MATERIALS
- Mixed-Up Words reproducible (page 28)
- dictionaries (optional)

OBJECTIVES

Students will
- spell words correctly.
- place words in alphabetical order.

Brainstorm with the class nutrition vocabulary words. Record student responses on the board. Give each student a Mixed-Up Words reproducible. Tell students that the words on the reproducible are similar to the nutrition words they brainstormed but these words are all mixed-up. Explain that students will unscramble the mixed-up words and then draw a line to the match. For example, a student would draw a line from *umcilac* to *calcium*. Once students have matched all the words, have them place the words in alphabetical order at the bottom of their paper. To extend the activity, have students choose five words from the list and write the definition of each word.

Food Guide Pyramid

The Food Guide Pyramid is an outline of how much food to eat from each food group each day.

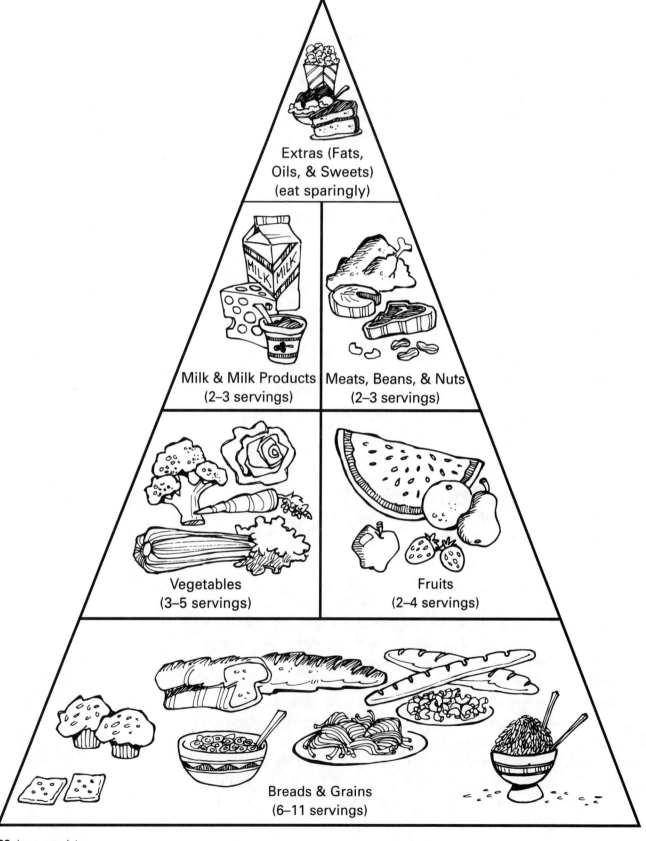

 # Steps to Digestion

Nutrients enter the bloodstream through the small intestine's lining.

The small intestine takes the thick liquid and absorbs the nutrients through its lining.

The large intestine absorbs water and passes on the unused parts.

The esophagus in your throat pushes the food down to your stomach.

Then, the larger intestine takes over.

Your stomach mixes the food with acid until the food is a thick liquid.

You grind the food in your mouth and mix it with your saliva.

Names _____ Date _____

 # Food Label Scavenger Hunt

Find a food that has . . .

0 grams of total fat _____

sugar as the first ingredient _____

100 calories or less _____

at least 5% of the recommended daily allowance of vitamin C _____

over 250 milligrams of sodium _____

10 milligrams or less of cholesterol _____

30 grams or less of total carbohydrate _____

more than 15 grams of total fat _____

more than 2% of the recommended daily allowance of iron _____

Watch What You Eat

Name_____   Date_____

Five Senses Poetry

Directions: Use your five senses to describe your food.

A(n) . . .
(name of food)

looks _____

smells _____

tastes _____

feels _____

sounds _____

Nutritious Pie Recipe

Directions: Create a nutritious recipe that will give you sweet dreams!

From the Kitchen of

_____

Ingredients

_____   _____

_____   _____

_____   _____

_____   _____

Directions

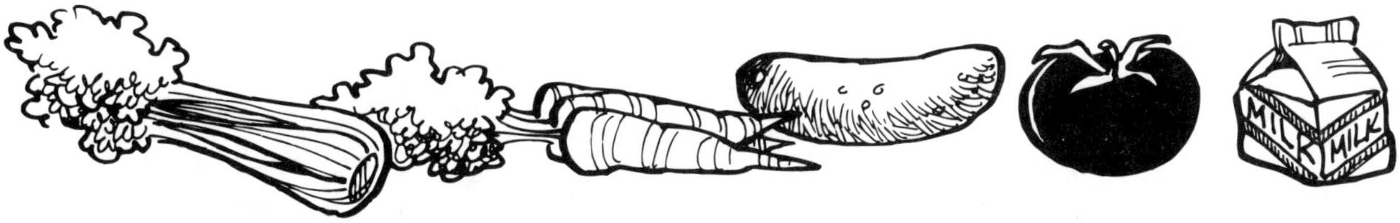

_____

_____

_____

_____

_____

_____

_____

_____

Name_____ Date_____

Mixed-Up Words

Directions: The following nutrition words are all mixed-up. Unscramble each word and draw a line from the unscrambled word to its match. Then, write the words in alphabetical order.

Mixed-Up Words	Unscrambled Words
umcilac	bread
matiinv	mineral
ropeitn	milk
alinerm	energy
earbd	fruit
ftiur	calcium
gerney	vitamin
kilm	protein

Words in Alphabetical Order

_____   _____

_____   _____

_____   _____

28 Language Arts

Math

Fruity Favorites

OBJECTIVES

Students will
- record numerical data in systematic ways.
- answer simple questions related to data representations.

MATERIALS
- Fruity Favorites Bar Graph (page 36)
- overhead projector/transparency
- variety of fruits
- knife (teacher use only) (optional)
- paper cups (optional)
- forks (optional)

Ask each student to bring his or her favorite fruit to school. Bring additional fruit to class for students who do not bring their own. Copy the Fruity Favorites Bar Graph onto an overhead transparency, and copy a class set. Explain to the class that fruits are an important part of a healthy diet. Tell students that fruits are a good source of vitamin A, vitamin C, and fiber. Explain that vitamins A and C help keep skin and tissue healthy and fiber helps in the digestion of food. Place the fruits on a table in the front of the classroom. Display the overhead transparency, and give each student a Fruity Favorites Bar Graph. Tell students that they are going to graph the number of fruits. Explain that students should shade in one box for every two pieces of fruit in a specific category, starting with the box closest to the label. Invite a volunteer to select a fruit from the table and shade in the appropriate box on the transparency. Have students shade in the same box on their reproducible. Continue with the remaining pieces of fruit. Tell students that any fruit that does not fit into a specific category should be placed in the "other" category. Continue this process until students have graphed all the fruits. Then, have them answer the questions at the bottom of their paper. Invite volunteers to share their answers with the class. To extend the activity, cut the fruits into small pieces to make a fruit salad.

Marvelous Milk

MATERIALS
- Type of Milk reproducible (page 37)
- overhead transparency/projector
- various types of empty milk containers (e.g., whole milk, low fat milk, skim milk, chocolate milk, strawberry milk)

OBJECTIVES
Students will
- estimate, calculate, and solve problems involving addition and subtraction of two- and three-digit numbers.
- solve problems and justify their reasoning.

Have volunteers bring in various types of empty milk containers. Copy the Type of Milk reproducible onto an overhead transparency, and copy a class set. Display the different cartons of milk. Invite volunteers to identify each type. Explain that today there are many varieties and flavors of milk and each type provides the same kinds of nutrients. However, point out that milk varies in the number of calories it provides and the amount of fat it has. Explain to the class that a *calorie* is a unit of measure used to describe how much energy a person receives from food. Tell students that although people need calories for energy, the body can use only a certain amount at a time. Calories that the body does not use are stored as fat. Tell students that taking in too much fat and calories can cause weight gain and health problems. Invite volunteers to read the nutrition label on each milk container to find out how many calories there are in one serving. Also, have volunteers figure out how many calories are from fat. For example, one serving of 1% low fat milk may contain 120 calories. Of the 120 calories, only 20 calories come from fat. Display the transparency, and give each student a Type of Milk reproducible. As a class, complete the table according to the class' findings, and then have students work with a partner to answer the questions at the bottom of their paper.

How Much in a Can?

OBJECTIVES

Students will
- know that when all fractional parts are included, such as four-fourths, the result is equal to the whole and to one.
- compare fractions represented by drawings or concrete materials to show equivalency and to add and subtract simple fractions in context.

MATERIALS
- Measure It reproducible (page 38)
- assortment of canned fruits and vegetables that show servings in ¼-, ⅓-, ½-, and 1-cup measurements (include large and small cans)
- set of measuring cups

Divide the class into small groups. Give each group several cans of food in various sizes. Have the group members read the nutrition information on the labels. Remind students that in addition to a list of ingredients and nutrients, each food label includes information about the size of one serving and the total number of servings in the can. Have students determine that servings for fruits and vegetables are often measured in cups. Hold up ¼-, ⅓-, ½-, and 1-cup measuring cups, and point out the different sizes. Tell students that they are going to be "food detectives" and that they need to determine how many cups of fruits or vegetables are in each can by reading each food's nutrition label. For example, if a can of pineapple contains four servings, and one serving equals ½ cup, then draw on the board four small bowls representing four ½-cup servings. Guide the class to see that two ½-cup servings equal 1 cup, that four ¼-cup servings equal 1 cup, and three ⅓-cup servings equal 1 cup. Give each group a Measure It reproducible. Encourage group members to analyze their food labels, record their findings on their reproducible, and then answer the questions at the bottom of their paper.

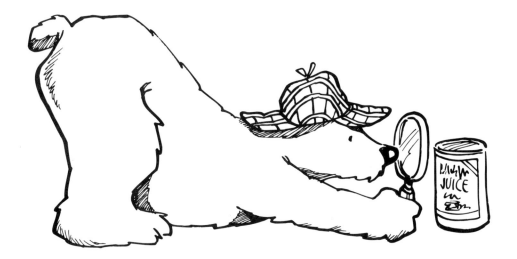

MATERIALS

- Food Guide Pyramid reproducible (page 22)
- grocery store flyers

Grocery Store Challenge

OBJECTIVE
- Students will model and solve problems by representing, adding, and subtracting amounts of money.

Divide the class into pairs. Give each pair of students a Food Guide Pyramid reproducible and a grocery store flyer. Tell students that they are going to pretend to plan their meals for a day. Have students use the Food Guide Pyramid to determine the number of foods from each group they need to purchase for a day's worth of meals. Ask students to look at their flyer and describe what they notice. Students should note that the flyer contains different types of foods and the price of each food. Have students use their flyer to determine the cost of each food item. Once partners have determined what food items to purchase, have them list the items by meal on a piece of paper. Have students write the cost of each food item next to the appropriate food. Then, have partners total the cost of their food items. Invite volunteers to share their food list and the total cost with the class. To extend the activity, have students graph the cost of each pair's list to determine the most expensive and least expensive lists. Challenge students to determine the difference in price between the least expensive and the most expensive list.

Terrific Trail Mix

OBJECTIVES

Students will
- identify and use appropriate units and measurement tools to quantify the properties of objects.
- measure items using ½- and 1-cup denominations.

MATERIALS
- Terrific Trail Mix reproducible (page 39)
- mixing bowls
- measuring cups
- mixing spoons
- raisins
- almonds
- O-shaped cereal
- shelled sunflower seeds
- resealable plastic bags

Tell students that it is important to eat healthy meals and snacks each day to help their bodies grow strong. Divide the class into small groups. Explain that group members will follow a recipe to make a healthy snack. Give each group a Terrific Trail Mix reproducible, a medium-sized mixing bowl, measuring cups, a mixing spoon, raisins, almonds, O-shaped cereal, shelled sunflower seeds, and a resealable plastic bag for each group member. Have groups follow the directions on the recipe to make their tasty treat.

Math 33

MATERIALS

- Calorie Chart (page 40)

Calorie Count

OBJECTIVES

Students will
- subtract three-digit numbers.
- collect numerical data.
- record, organize, and interpret data on a graph.

Explain to students that their bodies use calories for every activity they do, even reading and sleeping. Give each student a Calorie Chart, and explain that the chart lists some activities people do and the number of calories used by a person while doing the activity. Ask students what activity they could do for 2 hours that would use 280 calories. Ask them what activity they could do for 1 hour that would use 350 calories. Continue to have students determine different activities they would do to use a specific number of calories. Tell them to imagine that they just ate a meal with a total of 1,000 calories. Give each student a piece of paper, and have students determine what activities they can do and for how long to burn off the 1,000 calories from lunch. Have students share their results with the class. To extend this activity, invite them to track their movements for a 4-hour period of time. Have students determine the total number of calories used during that time frame and then record the calories and activities in a chart.

Food Pyramid Pizza

OBJECTIVES

Students will
- recognize and use fractions.
- add simple fractions.
- determine the duration or interval of time.

MATERIALS

- Food Pyramid Pizza reproducible (page 41)
- overhead projector/transparency
- paper plates
- English muffins
- toaster oven
- can opener
- canned tomato sauce
- measuring spoons
- aluminum foil
- shredded mozzarella cheese
- variety of chopped vegetables (e.g., tomatoes, onions, green peppers, squash, broccoli, green beans, mushrooms)
- dried oregano or basil

Invite volunteers to bring in and share with classmates one kind of chopped or sliced vegetable to be used to make a healthy pizza. Copy the Food Pyramid Pizza reproducible onto an overhead transparency, and display it. Read with the class the directions for making a healthy pizza. Give each student half of an English muffin and a piece of aluminum foil. Place the remaining ingredients on a table. Have students follow the recipe and "create" a pizza that includes ingredients from several food groups. Place each student's pizza on a piece of aluminum foil, and place it in a toaster oven until it is warm. Ask students to determine the amount of each ingredient they will need to convert the recipe for four pizzas. As an extension, ask students to create their own Food Pyramid Pizza recipe. Challenge students to include ingredients from all the food groups.

Name_____ Date_____

Fruity Favorites Bar Graph

	apples	oranges	bananas	pears	peaches	tangerines	other
20							
18							
16							
14							
12							
10							
8							
6							
4							
2							

How many fruits were brought to class?_____

How many students brought bananas and oranges? _____

Which fruit was brought by the greatest number of students? _____

Which fruit was brought by the least number of students?_____

How many more students chose the most popular fruit than chose

the least popular one?_____

36 Math

Name_____ Date_____

  # Type of Milk

Type of Milk	Calories in One Serving	Calories from Fat

Which kind of milk had the greatest number of calories? _____

Which kind of milk had the least number of calories? _____

Which kind of milk had the most sugar? _____

Which kind of milk had the least amount of sugar? _____

What was the difference in calories between a serving of whole milk and a serving of skim milk? _____

What was the difference in calories between a serving of 2% low fat milk and a serving of 1% low fat milk? _____

Which kind of milk had the greatest amount of fat? _____

Which kind of milk had the least amount of fat? _____

Which kind of milk is the healthiest for people to drink if they want to lose weight?

_____ Why?_____

_____

Names_____   Date_____

  **Measure It**

Directions: Work as "food detectives" to determine how many cups of fruits or vegetables are in each can. Then, answer the questions at the bottom of the page.

Type of Fruit or Vegetable	Number of Cups in the Can
_____	_____
_____	_____
_____	_____
_____	_____
_____	_____
_____	_____

Which can held the greatest amount of food?_____

How many cupfuls were in it? _____

Which serving size appeared most often? (for example: ½ cup, ⅓ cup, or ¼ cup) _____

Why do you think it is important to know the serving size and the number of servings in a can?

_____

_____

Terrific Trail Mix

Ingredients
1½ cups raisins
1 cup almonds
1 cup O-shaped cereal
1½ cups shelled sunflower seeds

Directions
Mix the ingredients in a bowl. Divide the mixture into plastic bags so you have an equal amount of trail mix for each group member. Try your tasty treat!

Calorie Chart

Approximate Number of Calories Used in an Hour	Activity
90	Little or no movement—breathing, digestion, reading, watching a movie, listening to your teacher, sleeping
140	Some movement—making a sandwich, getting dressed, taking a shower, brushing your teeth, walking slowly
200	Normal walking, making your bed, washing the car
275	Walking fast, roller skating, riding your bike, sweeping the floor
350	Playing soccer or baseball, jogging, swimming, skiing, running

Food Pyramid Pizza

Ingredients

½ English muffin

1 tablespoon tomato sauce

⅛ teaspoon dried oregano or basil

1 tablespoon plus 1 teaspoon shredded mozzarella cheese

3 teaspoons each of any chopped vegetables

Directions

1. Place half of an English muffin on a piece of aluminum foil.
2. Spread 1 tablespoon tomato sauce over the English muffin.
3. Sprinkle ⅛ teaspoon oregano or basil evenly over the tomato sauce.
4. Sprinkle 1 tablespoon shredded mozzarella cheese evenly over the tomato sauce.
5. Sprinkle 3 teaspoons of each vegetable of your choice over the cheese.
6. Sprinkle the remaining 1 teaspoon mozzarella evenly over the vegetables.
7. Have your teacher place your pizza in the toaster oven until the cheese is melted.
8. Ask your teacher to remove your pizza from the toaster oven and place it on a paper plate.
9. Let the pizza cool.
10. Enjoy!

Science

MATERIALS

- *Good Enough to Eat: A Kid's Guide to Food and Nutrition* by Lizzy Rockwell
- Strong Bones reproducible (page 50)
- clean, thin, uncooked chicken bone (e.g., thigh bone, wing bone)
- jar with a lid
- vinegar

Why Do I Need Them?

OBJECTIVE
- Students will predict the outcome of a simple investigation and compare the result with the prediction.

Read a few pages from *Good Enough to Eat* to the class. Invite volunteers to share things they learned about nutrition from the story. Record student responses on the board. Ask students what vitamins and minerals are and why it is important to take them. Review with students that vitamins and minerals are found in most of the foods people eat. If we eat a balanced diet, we will receive most or all of the vitamins and minerals we need. It is a good idea to take a multivitamin every day just in case we do not get all of our vitamins and minerals from the foods we eat. Explain the following information to the class:

- vitamin A—helps us see better
- the B vitamins—give us healthy skin and nerves
- vitamin C—helps prevent infections
- vitamin D and minerals such as calcium and phosphorus—are needed for strong teeth and bones
- iron—helps our blood carry oxygen

Invite students to predict what might happen if we do not get enough of certain vitamins or minerals. Tell students that they are going to conduct an experiment that shows what happens to strong bones when they do not get enough of certain vitamins and minerals. Give each student a Strong Bones reproducible. Share with the class a clean and dry chicken bone. Invite students to observe the bone and record on their reproducible what they see. Then, pass around the bone, and invite students to describe what the bone feels like. Explain to the class that you are going to place the bone in a jar of vinegar for five to seven days. Ask the class to record what they think will happen to the bone in the jar. Collect each student's reproducible, and redistribute the papers after five to seven days. Take the bone out of the jar, and rinse it with water. Pass the bone around the class, and have students record on their reproducible how the bone feels now. Invite students to compare the difference between the bone prior to placing it in the jar and after it is removed. Discuss with the class what might happen if the bone was left in the jar for 14 days and 30 days.

The Pyramid of Health

OBJECTIVES

Students will
- collect data in an investigation.
- analyze data to develop a logical conclusion.

MATERIALS
- Food Guide Pyramid transparency
- chart paper
- scissors
- overhead projector
- magazines or newspapers
- glue

In advance, draw a large bulletin board–size pyramid shape on chart paper, and cut it out. Divide the pyramid into six sections similar to the Food Guide Pyramid, and cut apart each section. Label each section, and set aside the pieces. Display the Food Guide Pyramid transparency. Ask volunteers to name the foods shown in each food group. Ask *Why do you think this is called a pyramid? Why is the bottom group larger than any of the other groups?* Explain that the base of the pyramid shows the foods we should eat most often. The top point of the pyramid shows the foods we should eat the least often. Discuss with students why the quantity of each group eaten is important to our health (e.g., too much sugar is bad for our teeth, grains contain good fiber). Divide the class into six groups. Give each group a section of the prepared Food Guide Pyramid. Ask groups to look through magazines and newspapers to find pictures of foods that represent their section of the food pyramid. Have students cut out the pictures and glue them on their paper. Reassemble the pyramid, and display it. To extend the activity, make a class tally of the types of food each student eats for lunch on one particular day. Post a tally chart on the board that lists the six food groups in the left column. Have students raise the number of fingers that shows how many servings they have in their lunch (e.g., if a student has a banana and a cup of applesauce, this would count as two tallies on the chart). After tallying all the foods, have students analyze the data and tell which food group is consumed the most and which is consumed the least. Have students compare their own lunch to the recommended number of servings per day.

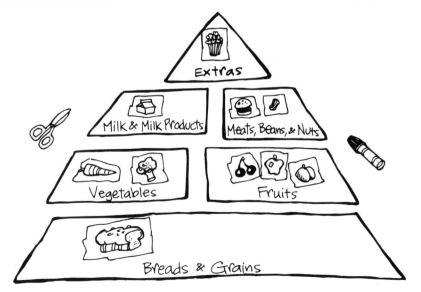

Water for Life

MATERIALS
- masking tape
- plastic wrap

OBJECTIVE
- Students will predict the outcome of a simple investigation and compare the results with the prediction.

Ask students to share what they think our bodies are mostly made of. Share with the class that water is the most abundant nutrient in our bodies. Our bodies consist of up to 70% water. Water is in every cell and between every cell. Our blood, sweat, and tears are all made of mostly water. While we could survive for weeks without food, we can only survive without water for about 3 days. We need water because our bodies lose water all the time. Invite students to share some activities that cause their bodies to lose water (e.g., going to the bathroom, sweating, crying). Then, explain to the class that they are losing water through their skin even as they sit there. Divide the class into pairs. Give each pair of students some tape and a piece of plastic wrap. Remind students that they should not put the plastic wrap near their face. Ask one partner to put the plastic wrap around the hand of the other partner and wrap tape around his or her wrist to secure it. Remind students that the tape should be snug but not tight. Have them wait 10 minutes. Then, ask them to check the plastic wrap. Discuss with the class what they observed. Students should determine that water is beginning to condense on the inside of the plastic. Point out that their skin is giving off water and the plastic traps it. Ask students *What caused your hand to sweat? How much water do you think a person needs each day to replenish the lost fluid and to stay healthy?* Ask students to determine what would happen if they placed plastic wrap around their foot. Ask *Would your foot give off more or less sweat than your hand?*

Collecting Water

MATERIALS
- chart paper
- garden trowel or shovel
- small cup
- plastic wrap
- small rocks
- marble

OBJECTIVES
Students will
- discover that evaporation and melting are changes that occur when objects are heated.
- learn that rock, water, plants, and soil provide many resources.
- follow oral directions for a scientific investigation.

Explain to the class that it is important to drink plenty of water every day to help keep our bodies healthy and hydrated. Have the class brainstorm different locations they might find water. Ask students to predict what will happen if a cup is set in dirt and covered with plastic wrap. Record student predictions on a piece of chart paper. Tell students that they are going to complete an experiment to see if they can collect water where they think there is none. Find a spot of dirt outside of the classroom that is in the sun for part of the day. Invite volunteers to use a trowel or shovel to dig a hole that is twice as deep and wide as a cup. Have a volunteer set a cup in the bottom of the hole. Ask another volunteer to cover the hole with plastic wrap and secure the edges of the plastic with small rocks. Invite a volunteer to gently push down on the plastic wrap and place a marble in the center to create a V-shape or depression. The marble should be over the cup in the hole. Have the class check both the plastic and the cup every other day. Record on a piece of paper any changes that appear. Within 24 hours, the cup should contain water. Have students record any changes in a journal. Students should discover that the water evaporates from the soil, is collected by the plastic wrap, condenses, and runs down into the cup.

Science 45

The Water in Food

MATERIALS

- paper plates
- slices of banana, orange, carrot, and cheese
- bread
- lettuce
- balance scales and weights
- drawing paper
- chart paper

OBJECTIVES

Students will
- collect data in an investigation.
- predict the outcome of a simple investigation and compare the result with the prediction.
- compare common objects according to physical attributes.

Invite students to name some drinks (e.g., juices, soft drinks, milk). Tell the class that all drinks contain water. Explain to the class that we also get water from the foods we eat. Some foods contain lots of water, and some contain very little. Divide the class into small groups. Give each group a paper plate with a banana slice, an orange slice, a piece of bread, a leaf of lettuce, a carrot slice, and a balance scale and weights. Give each group a piece of drawing paper. Invite groups to draw three columns on their paper and label the first column *Food Item,* the second column *Beginning Weight,* and the third column *Ending Weight.* Ask groups to list the food items in the first column. Then, have students weigh their food items and record the weight in the second column. Once all the food items are weighed, have them place the foods back on the plate and leave the items out in a dry, warm location. Invite students to predict which food has the most water to lose and which has the least. Record student responses on a piece of chart paper. After a few days, have students reweigh each food item and record the weight in the third column on their paper. Have them compare the new weight to the starting weight. Discuss with students why some food items lost more weight than others. Explain that the more water there is in the food, the greater the weight change will be. Ask *Which food lost the most water? Which food lost the least amount?*

46 Science

Fats in Foods

OBJECTIVES

Students will
- discover that sources of stored energy take many forms, such as food.
- collect data in an investigation.
- follow oral instructions for a scientific investigation.

MATERIALS
- paper bags
- scissors
- paper plates
- slices of apple, cheese, avocado, and banana
- mayonnaise
- plastic spoons
- eyedropper
- water

In advance, cut a paper bag into six 3" (7.5 cm) squares for each group of students. For each group of students, place a slice of each of the following foods on a paper plate: apple, cheese, avocado, and banana. Then, place a small spoonful of mayonnaise on the plate. Set the prepared plates aside. Explain that some fats and oils are important to our bodies. They provide stored energy, transport some vitamins, keep our skin healthy, insulate us from the cold, and protect our bodies from injury. However, too much fat and oil can be unhealthy. While it is easy to see that some foods, like cooking oil and butter, contain fat, other fatty foods are not as easy to identify. Divide the class into small groups. Give each group a prepared paper plate, six paper bag squares, and a plastic spoon. Model for the class how to write the name of a food item at the bottom of a paper square. Then, demonstrate how to smear that item on the paper square. Have students label their paper squares with the names of their food items and write *Water* on their remaining paper square. Have them smear each food on a paper square to leave a thin film and throw away any excess. Use an eyedropper to place a drop of water on each group's last paper square. Ask students to predict which foods will contain oil and which will not. Ask students to observe and compare the squares. Have students place their paper squares in a sunny location to dry. Explain that any water will evaporate but any oil will remain. The remaining oil will look "wet." After 30 minutes, have students examine their paper squares. Ask *Which foods contained oil? Which did not? Were you surprised by the results of any of the foods?* Ask students to discuss whether their predictions were correct.

MATERIALS

- Watch What Happens reproducible (page 51)
- small bowls
- water
- teaspoons
- cooking oil
- dishwashing detergent (Dawn™ is recommended)

Breaking Apart Fats and Oils

OBJECTIVE
- Students will follow oral instructions for a scientific investigation.

Ask students why they think fats and oils are important. Then, explain to students that one reason they are important is that they provide stored energy. Our bodies break the fats and oils down into very small globules that are then digested. Explain that this happens in the small intestine where a liquid called *bile* breaks down the fats and oils. Divide the class into small groups. Give each student a Watch What Happens reproducible. Tell students they will answer the questions on the reproducible as they complete the experiment. Give each group a small bowl half filled with water and a teaspoon. Ask students to predict what will happen when they add cooking oil to their water. Tell students to record their prediction on their reproducible. Pass around the oil. Have each group add 1 teaspoon of cooking oil to the water and watch what happens. Make sure students do not stir the oil and water. Encourage students to record the results on their reproducible. Then, ask each group to carefully stir the oil and water. Ask them to observe what happens after stirring and again after 5 minutes. Tell students to record their finding. Explain that their bodies need to break down the oil into small globules before the energy from the oil can be used. The dishwashing detergent will act on the oil like bile does in their own bodies. Have each group squeeze a drop of detergent into the oil in the water. Ask *What happened when the detergent hit the oil?* Then, ask students to stir in the detergent and observe what happens right away and after 5 minutes. Have students record their results. Ask *How was this different than when the mixture was only water and oil?* Encourage the class to discuss what they learned about fats and oils from this experiment.

Vitamin C Is Good for Me

OBJECTIVE
- Students will collect data in an investigation and analyze the data to develop a logical conclusion.

MATERIALS
- paper plates
- lemon and orange slices
- apple
- banana
- knife (teacher use only)

Explain to the class that vitamins are important substances that our bodies need for growth and to function properly. Most of the vitamins we need we must get from the foods we eat. Vitamin C is one of these vitamins. Write on the board *antioxidant* and *oxidation*. Explain to the class that vitamin C is important because it is an antioxidant, which means it decreases or stops oxidation. Tell students that oxidation happens when cells are damaged and exposed to air. This is most easy to see when we cut fruit. Over time, fruit begins to turn brown because the exposed cells are being oxidized. Vitamin C can slow down or stop this browning. Divide the class into small groups. Give each group of students three paper plates. Ask groups to label one plate *Lemon* and one plate *Orange* and leave the remaining plate blank. Explain to the class that each group will receive three slices of banana and three slices of apple in addition to a slice of lemon and a slice of orange. Tell students they are to place the lemon slice and orange slice on the plate with the matching label. Then, when they get the apple and banana slices, they need to place one apple and one banana slice on each plate. Explain to the class that as soon as they get their fruit slices they need to immediately squeeze the lemon juice over the fruit on the "Lemon" plate, and orange juice over the fruit on the "Orange" plate and leave the fruit on the blank plate alone. Cut the fruit, and immediately pass out the slices to each group. Ask students to observe and compare the fruit on each plate. Invite group members to share their results with the class and review the importance of vitamin C in our diet. Students should notice that the fruit without the lemon or orange has started to turn brown and oxidation has occurred.

Science 49

Name_____ Date_____

Strong Bones

1 How did the bone look at the beginning of the experiment?

_____

2 Describe how the bone felt at the beginning of the experiment.

_____

3 Predict what you think will happen to the bone after it is in the vinegar for 5–7 days.

_____
_____

4 After the bone was removed from the vinegar, how did it feel?

_____

5 Why do you think the bone feels this way?

_____
_____
_____
_____

Name_____   Date_____

Watch What Happens

1 Predict what will happen when you add cooking oil to the bowl of water.

_____

2 Describe what happened when you added the oil to the water.

_____

3 Describe what happened when you stirred the oil and the water.

_____

4 What changes, if any, occurred after 5 minutes?

_____

5 Describe what happened when the detergent hit the water and oil. Explain why you think this happened.

_____

_____

Science 51

Social Studies

MATERIALS

- *Long Ago and Today* by Rozanne Lanczak Williams
- Past and Present reproducible (page 58)
- overhead projector/transparency

Comparing Kitchens

OBJECTIVE

- Students will understand how limits on resources affect consumption of foods.

Copy the Past and Present reproducible onto an overhead transparency. Ask volunteers to brainstorm foods and cooking tools they think may have been in a typical kitchen in the early 1900s and how families obtained basic foods, such as milk, butter, eggs, meat, and vegetables. Record student responses on the chalkboard. Read *Long Ago and Today* to the class, and discuss the differences of the past and present. Display the transparency. Invite volunteers to compare and contrast at least three similarities and differences between modern kitchens and the kitchens people had in the early 1900s.

MATERIALS

- National Nutrition game board (page 59)
- crayons or markers
- laminating materials (optional)
- dice
- counters (e.g., beans, coins, circular chips)

National Nutrition

OBJECTIVE

- Students will gain a better understanding of the government's role in agriculture.

Make several copies of the National Nutrition game board, color the game boards and, as an option, laminate them. Explain to students that the government is directly involved in agriculture. Tell students that *agriculture* refers to preparing soil, producing crops, and raising livestock. Ask students why they think the government is involved with the way that food is grown. Prompt them to say that the government makes sure that the food grown on farms is safe for us to eat and helps farmers when disasters, such as droughts or floods, ruin their crops. Divide the class into pairs. Give each pair of students a prepared game board, a die, and two counters. Tell students to place their counter on "Start." Have players take turns rolling the die and moving the corresponding number of spaces. Tell students to discuss the captions as they play the game and to follow the directions when indicated. Tell students the object of the game is to be the first player to reach the finish line. To extend the activity, have each player write three things he or she learned about nutrition from playing the game. Invite volunteers to share their paper with the class.

Graphing What to Grow

OBJECTIVES

Students will
- understand that individual economic choices involve trade-offs and the evaluation of the benefits.
- analyze data to determine which crop most students would like to eat.

MATERIALS

- Graphing What to Grow reproducible (page 60)
- crayons or markers

Brainstorm with the class different types of foods that farmers grow (e.g., corn, strawberries). Record student responses on the board. Ask students to determine if the foods listed on the board are part of a healthy diet. Divide the class into groups of three or four students. Give each group a Graphing What to Grow reproducible. Ask students to decide what four different crops they would like to grow if they owned a farm. Have groups write one crop in each of the spaces along the bottom of the graph. Invite a group to read their four crops to the class. Ask the group to reread their crops, and invite students to raise their hands when the crop they like to eat best is named. Have the group color on their graph the number of boxes that matches the number of students who like each crop. Invite the remaining groups to repeat the process. Then, ask groups to analyze their data to determine which crop most students would like to eat. Tell students to complete the sentence at the bottom of their graph. Use the graphs to discuss how people who buy or consume goods affect how much is produced.

Social Studies 53

MATERIALS

- *From Cow to Ice Cream: A Photo Essay* by Bertram T. Knight
- Good to Grow reproducible (page 61)
- variety of maps
- crayons or markers
- reference materials (optional)

Good to Grow

OBJECTIVE
- Students will use a topographical map to determine the best location to grow crops.

Read *From Cow to Ice Cream* to the class. Discuss the process ice cream goes through. Tell students that the same is true for all types of crops. Brainstorm with the class what a crop needs to grow. Lead students to say *Crops need plenty of sunshine, fertilization, and water to grow well.* Ask students what negative factors might get in the way of a crop growing. Lead students to say *If the temperature is too cold, if there is a frost or snow, or if the soil or weather is too dry, crops will have trouble growing.* Share with the class different maps. Review with students how to read a map and a legend. Give each student a Good to Grow reproducible. Tell students they are going to determine where the best place would be to establish a farm and plant some crops. Encourage students to think about the factors that contribute to crops doing well and the negative factors that hinder growth. Discuss with the class how to read the key on the map. Invite students to draw a star to identify the place on the map they would establish a farm and then color their map. Encourage students to explain at the bottom of their reproducible what geographical features influenced their choice of location. To extend the activity, challenge students to research what crops would grow well in the geographical area they chose and include them in their writing. Invite volunteers to share their map.

54 Social Studies

Wants and Needs

MATERIALS
- construction paper
- scissors
- glue
- crayons or markers

OBJECTIVE
- Students will discover the differences between a *want* and a *need*.

Create a sample school menu with healthy and unhealthy food choices. Make a copy of the menu for each student. Draw two columns on the board. Label the first column *Wants* and the second column *Needs*. Ask the class to share some things they want (e.g., a bike or a new toy). Record student responses in the "Wants" column. Ask students to share some things they need (e.g., food, water, and shelter). Record student responses in the "Needs" column. Explain to the class that there is a difference between wants and needs. Ask students to share the difference between the list of wants and the list of needs. Explain that the same is true for nutrition. There are some foods we want, such as chips and cookies, but we do not really need them for good health. There are other foods we need for good health, such as fruits and vegetables. Give each student a piece of construction paper. Have students fold their paper in half widthwise. Have students write *Wants* on the left half of their paper and *Needs* on the right half. Give each student a sample school menu. Have students look over the various foods on the menu. Next, invite students to cut out the food words that should go in the "Wants" section and glue them in the "Wants" column. Then, have students cut out the food words they need for good health and glue these words in the "Needs" column. Have students illustrate each food word on their paper and then share their completed paper with a partner and discuss why they placed each food item in each column.

Wants	Needs
bike	a home
Nintendo	food
scooter	water
	warm jacket

Social Studies

MATERIALS

- scrap paper

Food Election

OBJECTIVE
- Students will understand the voting process.

List on the chalkboard five or six different foods, such as cookies, pie, apples, carrots, and pasta. Make sure to include healthy and unhealthy food choices. Tell students they are going to vote for their favorite food on the list. Give each student a piece of scrap paper. Have students write down their choice and fold the paper in half. Collect the papers. Invite a volunteer to read the votes to the class. Use tally marks to record the votes next to each food on the board. Circle the most popular food item. Then, give each student a new piece of scrap paper. Tell students they will vote for the food on the list that they think is the healthiest. Repeat the procedure for collecting and tallying the votes. Draw a line under the food the class voted as the healthiest. Ask students to analyze why that food was chosen as the healthiest. Have the class compare the results between the most popular food and the healthiest food. Ask students *Was the favorite food also the healthiest food?* If students respond *no,* discuss why they think this is the case. Ask students to consider why their favorite food was not the healthiest food (e.g., cravings, taste, influence of advertising, eating habits, availability, fat content). Conclude by asking students to consider all these factors when choosing what to eat in the future.

Goods and Services

MATERIALS
- drawing paper
- crayons or markers

OBJECTIVE
- Students will understand that some businesses provide goods and others provide services.

Write *Goods* and *Services* on the board. Discuss with students how some businesses provide goods and some businesses provide services related to staying healthy, such as grocery stores, restaurants, and doctor's offices. Record student responses on the board. Give each student a piece of drawing paper. Tell students to choose one of the businesses on the board and draw a picture of it in the center of their paper. Then, have students draw smaller pictures around the building to represent the healthy goods or services provided by that type of business. For example, a student might draw a doctor's office, a doctor, and a patient feeling better. Have students label their business as a good or service. Display the completed papers on a bulletin board titled *Our Goods and Services Town*.

Social Studies **57**

Past and Present

Kitchens of the Early 1900s	Kitchens Today

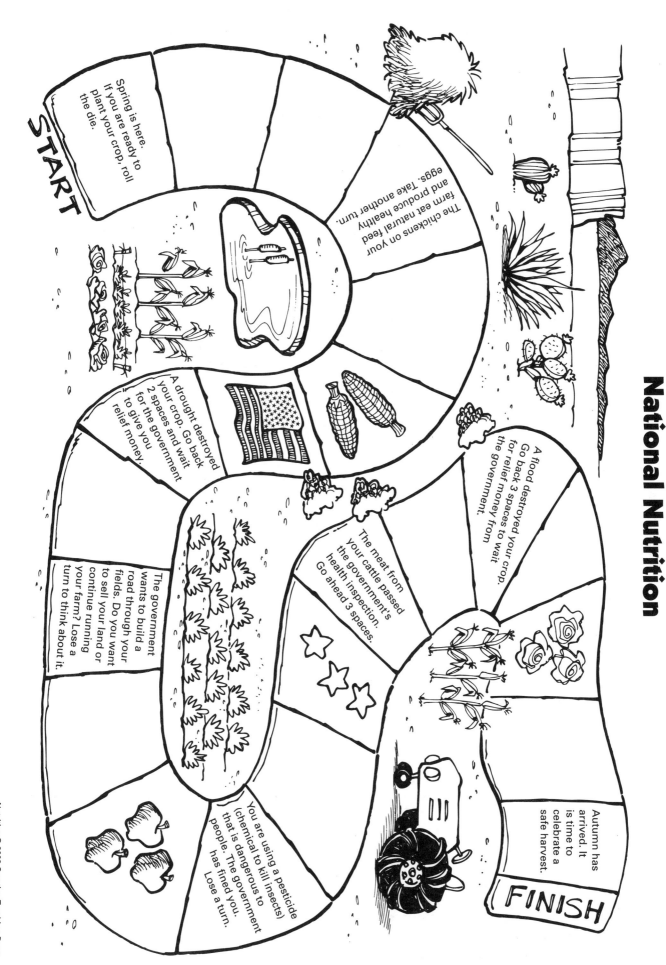

Names | **Graphing What to Grow** | Date

20
18
16
14
12
10
8
6
4
2
0

Our farm will grow a crop of _____ because _____.

Nutrition © 2003 Creative Teaching Press

Name_____ Date_____

Good to Grow

I chose the starred area for my farm and crops because
_____
_____
_____.

Name_____ Date_____

Eating Right Cumulative Test

Directions: Fill in the best answer for each question.

1 Which of these groups is **not** part of the Food Guide Pyramid?
- ⓐ Breads & Grains
- ⓑ Vegetables
- ⓒ Candy
- ⓓ Milk & Milk Products

2 How many syllables are in the word **nutrition**?
- ⓐ 4
- ⓑ 3
- ⓒ 2
- ⓓ 1

3 Circle the correct punctuation mark for the following sentence.
Candy is so delicious
- ⓐ period (.)
- ⓑ question mark (?)
- ⓒ exclamation point (!)
- ⓓ comma (,)

4 Which of these words is **not** a noun (person, place, or thing)?
- ⓐ eat
- ⓑ cat
- ⓒ apple
- ⓓ house

5 Which of these is a nutritious food choice?
- ⓐ cookies
- ⓑ pasta
- ⓒ chocolate
- ⓓ chips

6 In addition to eating nutritious food, what else should you do to stay healthy?
- ⓐ Get plenty of sleep.
- ⓑ Drink plenty of water.
- ⓒ Exercise.
- ⓓ all of the above

62 Cumulative Test

Name_____   Date_____

Eating Right Cumulative Test

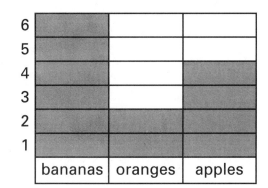

7 Look at the bar graph. How many more people like bananas than like apples?

ⓐ 2
ⓑ 3
ⓒ 4
ⓓ 5

8 Look at the bar graph. How many people participated in the survey?

ⓐ 8
ⓑ 10
ⓒ 12
ⓓ 14

9 How many cups are in a can of peaches that has four ½-cup servings?

ⓐ 1
ⓑ 2
ⓒ 3
ⓓ 4

10 If an apple costs 50 cents and a piece of pizza costs $1.50, how much money will both foods cost?

ⓐ $1.75
ⓑ $2.50
ⓒ $1.50
ⓓ $2.00

11 Why is it important to eat a variety of healthy foods from each food group in the Food Guide Pyramid?

_____

12 Describe a balanced meal you might eat.

_____

_____

Certificate of Completion

Congratulations!

_____
(Name)

You have become a Nutrition Nut!

_____
(signed)

_____
(date)